Sandro Rossetti

Press any key

O Mundo é um moinho

Angenor de Oliveira (Cartola)

A very special Thanks to PurpleGY
who translated and adapted this book
with love!

Chapter 1 - Eric

Eric had never understood many things that were clear and obvious for others. He had learned fast and well to ask himself questions to finally discover that was much more fun than finding answers.

Some mistook that for laziness. Others, for fear. For him, it was none of those things. It was a question of pleasure, nothing to do with rationality, calculation or convenience as an end in itself.

He might had found a way to exorcize his fears and create new ones. It had worked, at least to the point that he was really afraid.

Like every morning, Eric started his day preparing breakfast. His kitchen was full of devilries. Actually, others called them that way. To him, they were simple appliances. Eric had a passion for technology. Or at least he had been until a dangerous and devious virus affected him. They call it that way now. In the past, it would have been called 'the parasite of doubt', an annoying tick tock barging in his brain every time he realized to what an extent technology had changed his life and everyone else's.

Yet, this tick tock would not stop resonating in his head, a feeling of guilt, like the one you have sometimes after an orgasm. That instant, mild pain in the middle of the chest mixed with a sensation of ending.

Eric had no idea that, even those daily gestures that marked his breakfast with precision, would soon be distant memories.

Sarah, his partner, had also shared that passion. They met at work and liked each other instantly like two flowers blooming in unison, one in front of the other, even if they belong to different plants or, as he used to say romantically at that time, 'two terminals in an eternal handshake'. Being close to Eric had not been an easy task. She had known that since the beginning of their relationship. Too many ill-concealed doubts in one head, she thought. Too many questions that no one would answer. Supporting Eric in his difficult times was like moving a bath that was full to the top with water. Even if your movements had been soft, delicate, free from uncertainty, something would have probably got wet.

Their relationship was sailing smoothly until Eric started having problems at work.

Doubts about the real purpose of his work or at least about the actual effects it would have had on others. He had always been chosen to stand for other people's rights and he would even risk himself in person. Maybe with a little exhibitionism, he often repeated how his career choice had been dictated by his desire to create a better world through technology. He soon realized that a butterfly flapping its wings in Singapore would not always make a tree grow in Minnesota, and that the cause and effect chain may have a violent result.

Now, his world, just like everyone else's, was more virtual than real. The word socializing had become a synonym of social networks, and what used to be emotions and exchanges among people, had been reduced to simple emoticons or likes. Everything had to be expressed as briefly as possible. Was it because of Twitter? Was there really anyone to blame? He had studied a lot, but he would have never imagined that those communication protocols among computers would one day become the lexicon of humanity. He simply could not believe it.

Yes, he was certainly not the only one who asked himself these questions. Now, since 2018, there was a fairly consistent movement of individuals in the world who rejected totalitarian technology. That was why he had left his beloved appliances in the kitchen and, at the same time, closed all his social profiles.

He had read books, participated in conferences, become a convinced supporter of the movement and, of course, had quit his programming job.

Certainly, he could not deny the movement would use technology to communicate, but this small, imperceptible inconsistency had not affected him that much. After all, it was not about exorcizing technology but using it correctly or, at least, in the best possible way.

History, on the other hand, was a topic that had never betrayed him. He had already become a History enthusiast in the technical high school he attended, when he was bewildered by the professor's telling of past events. While he listened to past events, he asked himself thousands of questions about how people lived and communicated, and about the concept of travel, whose meaning has changed significantly over time. Living in a time when, at most, you would only travel a few hundred kilometers in your whole life? Unthinkable. He found it hard to imagine a previous world, without technology, without that little daily help that technology offers us.

Throughout history, men have used various media to communicate with posterity: books, scrolls, stone inscriptions. These objects had the difficult task to pass down stories, beliefs and all kinds of information. Some had worked. Others had been irremediably destroyed or were illegible such as the scrolls of Pompeii, which were intact but no longer legible.

What means would be used in his time to pass down knowledge for posterity? USB flash drives?

Stories of fires, raids, catastrophes that could have erased everything but, fortunately, had left something safe. That something would allow future generations to know, comprehend, study and learn from other people's mistakes.

He was astonished by the phrase 'history repeats itself'. A few years later, the Propellerheads would also release a song with this title: 'History repeating itself', which, at the same time, was a cover of an old song. Was music repeating itself, just like history?

That fascinated him and, even with the regular precautions, he was convinced that that phrase was perfectly aligned with the undulating essence of the materials that surrounded him. Eric had studied voraciously the undulatory theory of Mechanics, which says we are made of waves, waves of energy that, as is well known, are neither created nor destroyed but transformed.

Whether it was a Scientific American article about the theory of unification of the fundamental forces of nature, Star Wars and the theory of the Force or a book about Reiki, for him, everything was perfectly plausible or even obvious... Clear and obvious, while so many other things in his life were not.

Chapter 2 - Swan

Swan never agreed on anything. It was as if she needed confrontation to understand others or herself. Her body was slender but muscular. If you looked at her properly, even if her femininity sprang up from every pore, she could easily be transformed into a man by covering the right parts. On a motorbike, with her helmet, especially during the summer, some had mistaken her for a man.

She had met Eric at university, certainly not in the same faculty. Swan had attended "unsuccessful Philosophy", as she used to say. She had actually dropped out when she was one step away from graduating. Their encounter was actually, as it often happened with Swan, a confrontation. At that time, universities still held seminars, those encounters that were then replaced by direct communications in closed Facebook groups. Precisely that Facebook that years ago had become a global phenomenon which the United Nations decided to nationalize or, shall we say, 'turn into a global resource with the sponsorship of the United Nations' directly at the address fb://. Some qualified this as a communist expropriation, others, as an act of planetary security. Many others thought about a (more or less fair) exchange. However, at that time, seminars were still held live in a classroom at university, and the subsequent sessions of Q&A would sometimes turn into a physical confrontation among the participants. In other words, things from a different world.

An engineer-to-be and a future philosopher arguing about the topic of the seminar: 'The role of history, and questions without answers.' They had held it for 20 minutes, during which Swan and Eric had argued about the role of history and its effects on future generations. Swan stated that history did not repeat itself, that it was a dumb simplification for simple minds or, even worse, just the need to give an answer even when there is no answer. Eric's opinion, on the other hand, was exactly the opposite. Swan was a woman with her feet on the ground, while Eric was a dreamer who was ready to believe even in the impossible if it made sense. Logically, the dreamer should be the philosopher, and the rational one, the engineer. But they were different, entirely different from the others. The moderator had more or less calmed them, kindly inviting them to leave and continue with their confrontation outside because they were out of time, and it was time for another seminar, with a terrifying name: 'Religion: opium or purge of the people?'.

When they heard the moderator say the title, Swan and Eric first got quiet in unison, and then laughed out loud interminably, without realizing that the audience had already changed, and the spectators were sitting there precisely to attend that seminar.

They left together, still laughing, and a few minutes later they were inexorably together. They would be for the rest of their lives.

That was how their friendship started, a friendship with everything but sex, at least with the direct participation of both of them. Swan, in fact, had stolen a couple of girlfriends from Eric, but this had not undermined their relationship in any way. In fact, it had strengthened it. On his part, Eric had also made a couple of trips with Swan's partners.

They loved this secret axis between two beings from such distant faculties. They felt a bit like two infiltrated viruses, two organisms hosting a world that did not convince them anymore. They felt it hostile and, especially, unsafe. They shared a passion for history, they loved the same movies but, above all, they respected each other. They were Harry and Sally without the emotional ending, said Swan. Two tuning forks in resonance, according to Eric.

Eric and Swan lost track of each other for almost ten years after college, but after that, thanks to technology, they met, saw and hugged each other again, and discovered each other's things one more time. After dropping out from university, Swan had also started working on software development, which was already a common destiny for many people. After all, writing software is like teaching someone stupid but willful many simple tasks that must be done very quickly.

Swan had fun testing others' intelligence with tricks and fake questions. She was incorrigible. She liked provoking, not only sexually, but also in every other possible sense. She loved making up absurd stories to understand when the other person would rise, open his arms and discover the deception. Often, this did not happen. Swan used to say that men and women had lost a brain lobe. The impossible and the absurd were no longer contemplated.

She had accepted Eric's career decisions, and had listened patiently to all of his outbursts, questions and fears. It had been precisely Swan who had suggested Eric to quit his job, because he would probably find another one. He was too brilliant to fail.

'Quit,' she said, 'after all, you are doing something you don't give a damn about for people you can't stand'. Quitting due to 'dissatisfaction with the results of my work', is what Eric had written in his letter of resignation, even though the people who read it did not have the slightest idea of what it really meant. He had been a useless, grey bureaucrat at the software company where the person who read his letter of resignation worked. This person talked to Eric, tried to understand the reasons, and signed his resignation.

This man could not have known the disturbing truth Eric would discover a few years later.

Chapter 3 - 2020

For Sarah and Eric, 2019 had been a difficult year. Their relationship had been affected by Eric's work situation but, above all, by the different jobs each other had. Eric had started teaching Mathematics. He gave private lessons to make ends meet. That allowed him to survive and dedicate himself to his more or less secret passions. That's it, because Eric did not talk much to his friends about these things. He probably considered them immature to talk about certain topics. Sometimes he even wondered if that silence was a kind of fear, but sure about his certainties, he would have also cornered this in a place of his mind he called 'the bin'. This was where he kept all the unanswered questions and all the unquestioned answers.

Eric was a techie and he knew that, sooner or later, the bin would overflow. However, he kept filling it up without ever emptying it.

A Mathematics enthusiast like him, fascinated by numerology, could not but wait anxiously for 2020. It was a fascinating number, much more than the Mayans' dull 2012. Eric was very disappointed once 2012 ended and the world realized that the famous Mayan message did not have an illuminating meaning or a catastrophic consequence. More or less like the year 1000, when many expected the end of the world due to Christ's words 'The Thousand-Year revelation ', only to discover that the world would keep existing like before.

He and Sarah had also been to Mexico to visit the Mayan ruins. They had been to Chichén Itzá to admire the pyramids and, especially, the cenotes, subterraneous caves with fresh water that are interconnected by subterraneous canals. There are so many theories about the origin of the cenotes, including extraterrestrial theories, of course. For Eric, everything was pretty fascinating. When he came back home to Seattle, he started buying books on the topic, and attending interest groups and yoga lessons and, naturally, trying Peyote. The West Coast of the United States, from Seattle to San Francisco, was probably the best place to find people who thought like him. He had chosen to live there because of the movie 'The Birds'. In fact, he visited the city where the film was shot (Bodega Bay) and was fascinated, not much by the weather but by the light of the place. For Eric, the light was an extremely important part of life, and living in a place with adequate light would have a positive effect on both of them. The light, according to the undulatory theory, is a set of electromagnetic waves with variable frequency (from infrared to ultraviolet). Everything in this scope may be defined as light if it is perceived by the human eye. Everything left outside this scope is not light, at least looking at it with a human eye but, Eric said, this is only because we use our caliper to measure the world. Other frequencies may also be considered light, but they are invisible to us. When we say we are meteoropathic or we are happy lying in the sun, actually, we are only describing the effect of light on us, in our body or our mind. When we meet someone and instinctively have a positive feeling without making assumptions, maybe it is nothing but our response to what the other transmits to us. Mirror neurons work that way. Yes, precisely those exhaustively discussed neurons, capable of transmitting and receiving stimuli for actions among individuals. It is almost like saying that, if you see someone pee, you will probably remember you have to do it as well. Maybe, when two people love each other, it is nothing more than a mutual dribble, a form of energy coming and going between them and increasing gradually. It is almost like ping pong: if there is a feeling between the two players, the game goes much further than simply keeping score. It turns into a dance between two resonating human beings.

How many real-life situations can we compare to this? Many? All of them? Is everything a question of 'resonance'? Why are we moved with no reason? Why did Eric cry listening to 'Proud Mary'?

Theories about the undulatory nature of lights and its effects on the human mind had convinced him to settle in Seattle. It certainly was not Bodega Bay and the light did not have the same 'spirit', as he said. However, it was a good compromise for both their work-related needs.

In February 2020, a snowstorm flooded Seattle. It was an exceptional phenomenon, out of the ordinary. A thick layer of pure, compact snow had covered most of Seattle's territory, up to the Pacific coast. These were difficult days. Most essential services had been suspended: no water or gas, and the Internet worked intermittently, just like electricity. Ten days back to the Stone Age. A test for new generations who, surprised by the lack of means of communication, had no personal resources to survive to this isolation. Many experienced for the first time what it meant to live without mass communication tools. It was impossible to chat online, send messages or even know the weather forecast. Snow, which is basically water in the crystalline state, pure water, free from any mineral substance, was a perfect screen for mobile phone waves.

No telephone worked properly in the lower floors. Everything was covered by a thick layer of many meters of crystallized, pure fresh water.

There are people who accept reality as it is presented to them, without asking themselves too many questions. A few people look for a chair to stand up on it and see things from a different perspective. Eric did not. He knew that the carpe diem from the movie 'Dead Poets Society' had a precise and very solid basis. It was the key to open new doors, those doors that may finally provide plausible questions for the pending answers.

That is how Eric and Sarah were also locked up at home for days before the situation turned back to normal. Soon, the snow started to melt, and the surrounding green fields cheered up the neighbors again, as well as drinkable water, gas, the Internet and electricity. Very soon, those days of unreal white calm became a memory and the social networks were filled with pictures of an unforgettable Seattle. It seemed like everything had gone back to normal, without a trace of what had happened. The local administration worked very well. Seattle was a rich area, with one of the lowest unemployment rates in the world, and everything was working as usual again.

Nevertheless, an indelible trace had been left in Eric's mind; that feeling of emptiness again in his chest; once again it, imperceptible but present in every moment of the day, like a watermark in an unlicensed picture, which is there to remind you that you have not bought it and, beautiful as it is, you cannot use it. A mark he could not simply put in the bin. Why was this happening to him? Eric did not like the idea of having an extra preoccupation. He really did not need it. His brain already had a lot to do and his bin was now full.

Every day, he would get up, have breakfast, work, cook, talk, write, communicate, but nothing was the same anymore. He felt like a wife who had been cheated, capable of forgiving his husband's affair but not capable of forgetting about it. It was time to empty the bin.

Chapter 4 - Resonance

St. James Cathedral in Seattle was one of the oldest, most monumental and famous of the United States. The gothic shapes of the church reminded the inhabitants how strong the Pilgrim Fathers' legacy was. How much Europe had fertilized America with its culture and history.

If you really wanted a dreamy wedding, you had to book St. James Cathedral. Once a year, the magnificent bell towers of the church tolled on Saint James Day and flooded the surroundings with the sound of its gigantic bells. The sound was so loud that the neighboring building, the one in the intersection of Marion St. and 9th Avenue, who had been built many years before and was just in front of the bell towers, had to be modified during construction to prevent the strong vibrations from breaking the glass windows. In fact, the highest point in the building had a community terrace with a fully reinforced concrete wall, but it had a direct view of the cathedral from above.

A year after the disastrous snowstorm, the municipal administration developed an emergency plan and, aware as it was of service interruptions, installed huge screens on top of the buildings to warn the population in case of similar disasters. The idea was good, the modern version of fires lit on top of the hills to communicate, thought Eric. After all, it was a way to circumvent technology and still be able to communicate. One of the chosen buildings was precisely the one next to the cathedral. The panel was certainly not nice though. When switched off, they had managed to camouflage it with the gray of the building with a smoked glass. When the sun reflected on the glass surface. it was a different story but, deep down, it was bearable, said Eric, for the good of the community.

On Saint James Day, as was always the case in the last century, everyone met in the garden that surrounded the cathedral.

It was hot, but great trees provided some refreshment to the multitude gathered around the church.

At 12.00, as usual, the enormous bells started to move, and soon they were bombarding the whole city with the sound.

The neighbors were used to this uncomfortable reality. The previous day, they emptied their shelves, cabinets and anything that was at risk was firmly secured. They knew well the effects of low frequencies on buildings. Every inhabitant of this area was warned about the event a week before with a leaflet full of advice on how to make sure the sound of Saint James Cathedral's bells did not cause any damage.

However, even though they were used to this, they simply did not expect what happened that day after the fourth stroke. The sound of the bells, the left one in particular, which stroke fiercely against the nearby building in Marion St., had started to make the panel's enormous glass cover vibrate, a phenomenon known by every Physics student in the world: Resonance.

If an object capable of vibrating at a certain frequency is hit by a wave of the same frequency, it starts to vibrate, but the rest of them don't. It's like saying: I talk, but as long as nobody understands my language, my message does not have an effect.

Well, that day, the left bell tower's message certainly had a listener. The screen's glass, after some oscillations that were imperceptible at first sight, was shattered to pieces. Millions of pieces of glass were catapulted to the neighborhood, tiny pieces of shatter fell like rain over the church, the bell towers, trees, nearby buildings and, unfortunately, the crowd. The trees, capable of filtering sunlight but certainly not microscopic pieces of glass, were useless. It's hot in the summer, and the scarce light clothes did the rest.

A horrible experience: no one died but many were injured. It was as if a truck full of red paint had exploded: blood covered people almost completely, and then it covered things. This spectacle was recorded live and immediately made networks collapse. People screamed. Many had their eyes covered in blood and could not even remove it because their hands were covered in glass.

Saint James Day would never be the same, a bit like Eric after the snowstorm.

Chapter 5 - The Analysis

When he was just a child, Eric's parents gave him a work desk as a present, because they understood their son's ability for craftwork. Soon, they realized that craftwork was not an end in itself for little Eric, but a means to do other things. This do-it-yourself corner in the garage at his parent's house in Minnesota became, in a few years, a laboratory full of devices. From Chemistry to Physics, little Eric played with anything, without understanding everything completely. He was not persistent and would soon get tired and jump to something else, but not before learning the basic principles of each discipline. We are talking about the bare minimum, that minimum that allows you to have a conversation with those who probably know more without looking like a complete idiot. That was how, wandering among groups that fed on the energy of the universe and communities that proposed an alternative lifestyle, often inhabited by people with a glorious past in the working world, Eric discovered an enormous potential for knowledge.

That was how he had the opportunity to discuss with former teachers of Physics or Chemistry, biologists, computer experts, quantum physicists, astronomers, mathematicians or simply enthusiasts of a discipline. He often talked about the two previous years, what had happened to him, the disaster at Saint James and the record snowstorm of 2020. He was good at starting a story like a novel, without giving his listener any idea about his true purpose. He knew that, if he made direct questions, he would not have answers, only closures. Others had a sort of unconcealed distrust, as if everyone was up to something the whole time. Instead, if talking about this and that he slowly engaged his interlocutor, then the victim fell into the trap.

Here it was also a question of language, being able to become in resonance with another person trying to find the right frequency.

He talked about Saint James Day, not so much about the incident in itself (most people knew about this event), but about what we should learn about it. What had only been an inconvenience to the neighbors due to a design oversight had turned into a deadly trap for thousands of people. Despite the good intentions, they had not realized they had set a highly potential bomb to explode precisely on the most crowded day. Not even a jihadist would have been able to plan such an attack. However, a whole community did not think this could happen. Of course, there were movements against the installation for thousands of reasons, but no one would have seriously thought that it could blow up due to the sound. The same panels had been used in many other cities of the world: old cities, new cities, cities with entirely different buildings. However, this phenomenon had never occurred. Was it the classic necessary and sufficient condition, which was so difficult to his Math students? He had talked with some physicists about how it had happened, about why. How were these bells so different? What was unique in that sound? The answer was more or less the same every time: the right frequency with the right amount of energy, nothing more. Nothing that couldn't be reproduced theoretically in a lab. No witchcraft, astronomic or karmic events. Just a little bit of Physics, volume 1, and, of course, many coincidences.

Eric talked to a biologist about the role of trees in the disaster. Plants are filters. The air, soil, water and sunrays are filtered to different extents. However, tiny fragments of glass came intact through a dense network of leaves, branches, trunks... Was it only a question of measure? Yes, it was only that. Just like a Chinese strainer that strains rice properly but not couscous. His Electronics teacher was right: never say something is big or small without a reference. In other words: big or small... compared to what?

That was it. The glass fragments were small enough to pass through the trees. Those same trees that would had stopped much bigger objects or, simply, objects with different shapes, had become permeable to those tiny pieces of glass.

For months, he made questions and developed theories looking for something with no shape or name... Once again, he did not know whether he was looking for a question or an answer. However, one thing was certain. Eric was fascinated with both facts, probably for reasons he still could not verbalize, not even to himself, much less to others. He had something at hand. It was like a memory that becomes more and more present and clearer.

A kind of reverse déjà vu. Just like The Beetles records played backwards, the notes were the same, but the melody changed completely even though there was something familiar about it when you listened to it.

Chapter 6 - ELF

Swan lived in Madrid; she had moved there many years ago. She worked on the development of an NGO which was active in geological research in Africa. She communicated with Eric regularly, also because she was the only one who confronted him after Sarah left him in August 2024. Their breakup had been agreed and pacific, an adult 'this is as far as we go, no resentment'. Swan had been close to Eric and Sarah during their breakup and had managed to be Eric's best friend and a good friend of Sarah's. Swan's chameleonic body helped camouflage in the eyes of those who looked at her. She could direct her listener's attention to different parts of her body, as if she could control other people's framing. She could feel empathy for anyone whenever she wanted, a sort of palette where other people projected their idea of Swan, at her will. At the same time, you didn't need to ask if she didn't like you. She made herself understood immediately and clearly. In summary, a scorpion.

Not even her extremely pronounced nose had affected her. Not even Eric complaining with his truck driver humor that there was no more oxygen for him when he was in front of her.

Shale oil is a kind of fuel obtained through rock fragmentation after a series of steps, which are anything but harmless to the site that hosts the extraction. That is precisely what Swan did: identify the effects of very low frequency acoustic waves on the Earth's crust or most superficial layer. It was an interesting job, halfway between technique and research, and it allowed her to remain anchored to the disciplines she found more pleasant. Because, in fact, even though she was a technician, her job put Swan directly in contact with the local population, especially in Nigeria, one of the natural jewels of Africa, devastated by a wild search for oil and by the local government. In Nigeria, for years, there was an act of law punishing homosexuality with death by stoning. The subsequent liberation of the country through a coup in 2018 and the whole of next year, which had led to a kind of civil war, had exhausted the country, which was returning to democracy with difficulty. For Swan, it was the perfect place. Active in the recognition of everyone's rights, she quickly became an icon among the locals, especially on the coastal area, which was more open and modern.

Living in Europe and Africa gave her the opportunity to observe the two worlds and understand its peculiarities, defects and virtues. A point of view that had persuaded her of how many useless things we do thanks to technology and how many more we could have done if we had only turned off the device every now and then. Technology should have solved the small everyday problems and given us comfort and safety. However, it ended up being a weapon to limit ourselves and respond passively to life. That's what she thought in her heart. That's what she saw when she landed in Lagos and pressed the 'play' button of his second life again.

Her scientific work mixed with her social life. She was happy.

Eric called her many times. He told her he was trying to connect dots, dots of a general idea that was not clear to him yet. It wasn't easy for Eric to communicate with others, but, with Swan, a few words were enough to instill his thoughts.

They were like two computers that were connected to each other only a few minutes a day, and spent the rest of the hours developing calculations, tasks and prognoses separately. As Eric met people and talked to them, he communicated his impressions to Swan. Eric knew that, very often, unidirectional communications when Swan did not answer, meant she had nothing to say and that was all.

It was Swan who called Eric one night. She was in Lagos and did not mind if she woke Eric up in the middle of the night. Eric was sleeping and she left a message that said: Schumann Resonance.

In the morning, Eric read the message and started reading about this phenomenon. Schumann Resonance was predicted by mathematician Winfried Otto Schumann in 1952, but they managed to verify and measure it in 1963. It consists of general electromagnetic resonances excited by lightning discharges in the cavity formed by the Earth's surface and the ionosphere. In a few words, the ionosphere surrounding the Earth becomes like the lining on an electric cable, capable of maintaining extremely low-frequency waves which are inaudible for most equipment in resonance.

A frequency of 7Hz belongs to the extremely low frequency (ELF) spectrum, that goes from 3Hz to 30Hz. As every Physics student know, the higher the frequency, the shorter the wave. A common old FM radio works with a wavelength of just over one meter, to give us an idea. Now, what happens is that Schumann resonance frequencies involve very large electromagnetic waves (of approximately 50 000 kilometers). Under 3Hz, the wavelength is larger than 100 000 kilometers, but no instrument is capable of producing or identifying it, at least that was what was thought.

Swan was quite familiarized with the topic. Shale oil produced acoustic waves of various frequencies, including those around 10Hz. Sound waves, thought Swan, not electromagnetic waves, but this parallelism might make sense anyway. They were just different examples of a natural phenomenon and being able to look at an electromagnetic phenomenon with one's eyes (actually one's ears) was undoubtedly bold. Swan didn't care if her conclusions contrasted openly with science, logic and physics. Swan had realized that what both of Eric's experiences, the snowstorm and the Saint James Day tragedy, had in common was the wavelength. Snow can shield cell phone waves, and a glass fragment rain could probably do it as well. Trees could filter light, but they all let some through. What Swan had almost discovered was actually much more. As it was not possible to measure or generate very low frequency waves, nobody knew if there was anything that could block them.

Schumann resonance, which dealt with higher frequency waves, above 7 Hz, seemed to be the only physical phenomenon capable of making very low frequency waves resonate. Only the right lightning, phenomenon or instrument was missing to turn everything on. Was it possible that Mother Nature had not foreseen this? Could there be a kind of legal vacuum from 3Hz to zero?, Eric asked himself at that moment.

The answer was obviously NO, and it was there, right above their heads.

Chapter 7 - ...what if tomorrow...

History probably would not have occurred, and many discoveries would not have been made without the 'if'. 'If' is the precursor of the 'sliding doors'.

Those who study information technology know well the analysis of the formal correctness of an algorithm.

In principle, it is quite simple: 'If you must cut a lemon, make sure you can always cut it, whatever shape it has, in any situation'. This requires you to take with you every tool to cut the lemon and avoid, for example, having the knife and not the cutting board. What do you know about always having a place to cut at your disposal?

It may not be the best example, but it does transmit the idea quite well, even for a simple little head like mine. How can you predict something that cannot be seen or measure directly or indirectly? What would happen if an extremely low frequency became in resonance with the planet and, above all, what if it had already happened?

The sun was high in the sky of Lagos. It was hot but it was the solar radiation that was specially devastating. Such a high UV factor prevented everyone from going out without at least some protection. Those days, Swan was preparing reports about measurements made in the rocks of Nigeria. There were as many shale oil extraction side effects as you wished.

She was almost a Doctor of Philosophy. This "almost" had bothered her for her whole life. Sometimes it was a burden, sometimes a medal. However, William Shakespeare said that 'there are more things in heaven and earth than in the minds of philosophers.' Her chameleonic being had helped her in her life, but now it was about to give her something more. It would compensate her for all those 'almost' she had heard throughout her life.

Eric, in the meantime, in the midst of a brainstorm, had been reading everything about Schumann resonance since 7 am and, when Swan woke up, she found more than 10 messages from Eric. The right time to have some coffee and start: 'Hi, Eric. This is Swan'.

Saint James Day tragedy had changed society profoundly. The city of Seattle had faced something that neither economic comfort nor the best minds could have avoided. It was referred to as an unprecedented event. The gray glass protecting that panel, the only one that had been modified to improve its esthetic integration, had certainly made the difference but had not been the cause. Was it maybe the cutting board that was missing to cut the lemon? That lemon was special, and the algorithm to cut it had found a flaw, that was the harsh truth!

Eric answered the telephone immediately. Overcome by euphoria, he elaborated on everything he had learned about Schumann resonance and its effects. In summary, everything that Swan already knew, but listening to Eric had always been a source of inspiration for her. Recently awaken, she let Eric's baritone voice lull her, with her coffee still going through her stomach. Eric's last word was 'wavelength'. Then, silence.

Swan's brain was up and running and she said, almost distracted, 'you know, Eric, Seattle's trees are nothing compared to those in Nigeria'.

Once again, her Electronics teacher's lessons. Once again, big or small, but compared to what? Swan continued: 'Who knows what would have happened here. The trees may have stopped the debris, but solar radiation would have acted as a mirror to the great panel, and the fragment multiplying effect would have been replaced by thousands of flashes all over the city. Who knows'.

The low-frequency sound of the bells made the glass resonate. The subsequent burst made it explode like crystal glass breaks, with a very high-pitched sound. It's not necessary to bother Ella Fitzgerald, a tuning fork would suffice.

The long waves of a low-frequency sound had activated a multiplying mechanism and millions of tiny fragments were shot towards the crowd, trespassing trees, clothes, bags, backpacks, glasses and, many times, also the human body.

¿What if the same happened to Schumann resonance? What if the impossible was not such, but only 'unverified'? What would be the multiplying effect starting with long waves of more than 100,000 kilometers? Electromagnetic, not acoustic waves.

Chapter 8 - Fractals

Fractals had always fascinated Eric, and he was not the only one. Many people, regardless of their culture, education or social level, appreciate fractal shapes without knowing it. Those strange colors, often colored with a black background that outline a shape, are vaguely similar to the door of a Hindu temple. Often copied by Amsterdam coffee shops and the post hippie culture, fractals are enchanting only due to the almost hypnotic power they have even over the most distracted and skeptical observer.

Benoît Mandelbrot coined this term over 50 years ago, in 1975. According to the polish mathematician, 'it is believed that somehow fractals have a correspondence with the structure of the human mind, that is why people find them so familiar. This familiarity remains a mystery, which increases as the subject deepens'.

Eric also had beautiful framed picture of a fractal in his living room. After breaking up with Sarah, he had kept few things, and that painting was one of them. Actually, to appreciate fractals, you need a good sight or a computer to enlarge the image. That is how you discover that the main characteristic of these funny drawings is that the shapes are repeated in different scales, that is to say: when we blow up the image, we obtain a drawing that is similar, almost identic to the initial drawing and so on, to infinity.

With fractals, you never know what level you are at. Everything is the same and it is repeated to infinity.

It may have been precisely this characteristic that fascinated Eric. The fact that fractals were the quintessence teaching of his professor: big or small, but compared to what?

As usual, he had carried out as much research on the topic as he could and, in his wanderings on the West Coast, he discussed it with physicists, mathematicians, biologists and chaos theory scholars. In fact, fractals may be used to create mathematical models that describe natural phenomena, such as the development of a tree from the trunks to ever smaller branches. This applies to the fir tree in particular, but we can find fractals in the developments of fern leaves or in the shape of coastlines. The continuous and incessant erosion of the waves also made coastlines, including the one Eric was going along, developed according to a typical pattern of fractals, gulfs and smaller inlets, which are also part of other gulfs and bigger inlets.

Mother Nature does not do anything by accident, Eric knew it well. When he was asked if he was an atheist, he said no. When he was asked what god he believed in, he always answered 'Mother Nature, who else?'.

Since Swan had proposed her preposterous theory of Schumann resonance, he had spoken with everyone about it, asking whether it was somehow possible to create a mathematical model to simulate the effects of Schumann resonance at very low frequencies, under 3Hz. Almost everyone answered immediately that, in any case, that would not have been possible, that the absorbent properties of the ionosphere would have never let a wave of more than 100,000 km extend, that it was too big to be produced and measured.

Too big? What kind of answer was that? Too big as compared to what? To Earth, to the ionosphere's dimensions, to an atom, to what?

Eric was very annoyed by these answers. He always said it was useless to point at the moon if everyone looks at the finger. He talked about this with Swan for a long time. He felt frustrated by this isolating situation, he was at a standstill.

That was how Swan remembered his Mathematics professor at university. Yes, she had tried to attend Mathematics at university, but after the first test, professor Oelio had advised her to change to Philosophy. He told her that her mind, even though it was logical, was not meant for mathematical but for philosophical rigor. The building of the Faculty of Philosophy was next to the building of the Faculty of Mathematics and Swan, even though she was lazy, accepted professor Oelio's advice.

However, they remained friends. He was young and had just come from Brazil. He had a talent for teaching, more than many of his colleagues. He could express mathematical rigor without gimmick but, at the same time, without boring his students. Maybe due to his not very sculpted body, students found him adorable, extremely polite and with a melodic English, typical of those whose first language is Brazilian Portuguese.

Swan and Oelio remained in contact and, with time, mutual respect grew more and more.

'Talk to Oelio - suggested Swan - 'he is an open person, without prejudice, and crazy enough to listen to you until you finish'. Eric followed her advice and called him. Professor Oelio had already heard Swan talk about Eric, but it just so happens that they had never met or talked. It was like talking to an old friend that you had never met, a strange experience for both of them. Since their first exchange through the phone, Eric understood Swan's advice was sound. At that moment, Oelio was giving a series of conferences at universities in California. They met in Fresno, one of the ugliest places in the state, according to many. An urban agglomeration in California's backcountry, long perpendicular streets repeating themselves infinitely. Every corner of the city of Fresno is similar to all others, Eric used to say. Fresno was our representation of fractals, like saying: we managed to imitate Mother Nature very poorly!

Oelio was a pure mathematician. Obviously, he also had knowledge of Physics and Chemistry but, in his analysis, he never left the main road, the only true main road, free from external conditioning: Mathematics.

Oelio's hotel was made of wood and bricks, very inadequate for a place like that. It was reminiscent of the hotel in The Shining, but it was smaller, not due to its esthetic but due to the impact it made when you arrived by car in one of the few streets on a slope in Fresno. Eric got out of the car and saw Oelio's round profile. He was observing him behind the window, protected by the air conditioner. A handshake? A hug, exclaimed Oelio, who pounced on Eric almost as if he wanted to absorb him. Eric's slender body was surrounded by Oelio's. Eric did no less, he had forgotten how to greet people beyond the wall with Mexico.

Oelio let Eric talk during his whole exposition. He did not make a gesture or sign. He seemed to be distracted and constantly searching for food. He used to crunch everything, fill the table with crumbs and have fun breaking them into ever smaller pieces. When the crumbs were tiny, he put on his glasses and went on. He was fascinated by that infinite grinding process.

Eric told him everything, from the record snowstorm in Seattle to the tragedy at Saint James, ultralow frequency wave theories, Schumann resonance and the search for a new consequent mathematical model.

When he finished, there were 60 seconds of deafening silence. Eric's last words still resonated in both heads. Finally, Oelio started talking.

Chapter 9 - Divide et impera

'Swan did not lie', said Oelio with a mischievous, almost indecipherable smile, and his dark eyes slightly open. 'I think I get where you are going and, undoubtedly, it is a fascinating perspective'. Elaborating a mathematical model was certainly possible or at least worth trying. He said he would have to bother a friend who was a meteorologist and an astrologist. Eric jumped out of his seat when he heard the word 'astrologist'. 'Astronomer, you mean an astronomer, right?'. Oelio looked up and said 'No, I want an astrologist! Trust me, you need an open mind'. 'Yes but why an astrologist?' - insisted Eric. 'Because Mathematics is like bossa nova, not like techno'- answered Oelio - 'Have you ever listened to Jobim?' 'No' - answered Eric, overwhelmed by Oelio's penetrating gaze. 'But I will' - he concluded with his best smile; a smile he knew would exude all of his charm.

They kept talking. Eric tried to hide his racism towards astrologers for Swan's sake. They talked for another three hours. After that, a little tired, they greeted each other with the same enthusiasm they had at the beginning, or even more.

Two weeks went by. Eric got up, prepared breakfast, went to work, did his shopping, but he never stopped thinking about it. Every now and then, he checked Oelio's last access to the chat, but there was absolute emptiness for two weeks. He lived almost like an inmate. He had not even washed his clothes for two weeks. He had almost worn all of his wardrobe, work clothes, formal clothes and even cricket clothes. At three in the morning, Eric's phone started blinking, message after message from Oelio. Eric was sleeping. He left the telephone on but silenced the messages. After 10 minutes without an answer from Eric, Oelio decided to call him and wake him up. 'Hi, Eric. We have a mathematical model and the consequences are disastrous. The tragedy at Saint James is insignificant compared to this'.

Careless of his appearance, Eric left his house like a watchdog chasing the milkman, he got in his car and went where Oelio was. He lived two hours away by car. When he got there, it was dawning. The first sunrays hit the meadows and the surrounding countryside. Such white and pure light reminded him of the compact, solid white of the record snowstorm in Seattle. Oelio was wearing pajamas, he was sweaty and nervous but, when he opened the door of his house, he could not help but notice Eric's state. He had left his house in long pajama pants, a green terry cloth sock and a white cotton sock, a formal shirt, which was crooked and buttoned up wrongly, with cuff links, showing his abundant chest hair. He was wearing a dark blue raincoat which, along with a missing button in his pants, and the subsequent gap at the right height, he looked like a 'last-minute exhibitionist'. Oelio smirked, that sly smile again, probably more decipherable this time. Eric did not get very upset. He knew he had a certain influence over Oelio and that did not bother him much, in fact, it flattered him.

Oelio let him in and for a moment the hormones made him forget the reason for the visit, then the house mess made him return to reality. It was almost nighttime, but it seemed like midday at Oelio's house.

In the kitchen table, there were sheets of paper scattered everywhere, a couple of laptops, two cell phones, a tablet and an ashtray with numerous ends of smoked joints. Oelio offered him something to drink, eat and smoke, and started his presentation. He explained Eric that, even though it was just a mathematical model and it was not based on experimental evidence but just on theory, it had unimaginable consequences. He went towards a great painting he had in his living room and removed it from the wall, unveiling a board that looked like the ones in schools. They are very hard to find now, but Oelio had one installed in his house and covered it cunningly with a painting for esthetic reasons, although it was ready to reappear whenever he wanted. He said that he could not explain anything to anyone without a piece of chalk in his hand, including himself. Feeling the chalk between his fingers and writing on the board calmed him down and gave him confidence. With his didactic skills, and even though Eric knew something about differential equations, and after begging him to at least button his pants, Oelio explained Eric the model he had developed, the assumptions and the results of a simulation. If a Schumann resonance had been able to resonate in the ionosphere at less than 3Hz, it would have produced a phenomenon that, in its known form (between 7 and 30 Hz) could not happen. The generated waves would have divided in two components: one in the resonance frequency and the other one in a double frequency. This wave subdivision and frequency duplication phenomenon would have theoretically continued indefinitely. Actually, it would have disappeared because, each wave subdivision would have reduced half the energy and, as the frequency increased, the mountains, buildings, cars, objects and things would have absorbed part of the energy, each thing according to its own resonance frequency. Just like sliding a finger on a piano from left to right, a crescendo of unimaginable power waves. What made the difference was Schumann resonance, the fundamental one at a very low frequency would have worked as a carrier. Similar to what happens in frequency modulation radios: they transmit at a precise basic frequency and they also produce a second wave overlapping the first one, which contains the audio.

'A carrier below 3 Hz' - exclaimed Eric - 'Exactly' - said Oelio - 'a carrier, a radio signal, capable of going through any object, including a great part of the Earth's crust'. Eric still did not understand the reason for Oelio's scared look, he simply could not understand what those catastrophic consequences could be. We are bombarded with electromagnetic radiation every day without suffering any damage.

Electromagnetic radiation goes through our bodies every day. It occurs when we talk on the phone, when we undergo X-rays and, obviously, when we are under the sun. Some kinds of radiation may be harmful in the long term but, in general, hearing the term 'electromagnetic radiation' is not an alarm for human beings. On the contrary, if we were electronic equipment, we should be worried. Like having a pacemaker, electromagnetic radiation could be lethal for some devices.

Many years ago, Eric heard a high frequency technician claim he had listened to a Led Zeppelin song at full volume in the car radio with the car parked and the radio switched off under the antenna of a frequency modulation radio station. Of course, it was a 30-KW transmitter and the car was very close to it, but close to what?

I even talked about it with a meteorologist friend' -said Oelio, - "because Schumann resonances may be activated by lightning or intense electromagnetic radiation produced by the sun. Not only is it technically possible, but it is also extremely probable it has already happened'- concluded Oelio.

At that moment, Eric decided to smoke, drank a glass of water and said that, in any case, if it had probably occurred already, the more reason not to understand the catastrophic consequences of these very low frequency Schumann resonances and the subsequent escalation of secondary frequencies.

"The context" - exclaimed Oelio - "the context changes! You remember I told you I also wanted to ask an astrologist, don't you?' 'Sure'- answered Eric -'I listened to bossa nova like a fool because of you!' 'In one year, the Earth will complete a cycle around the sun. Without other references, this is the longest period we can measure, right? Without other planets, stars and other celestial bodies, we would not be able to measure any precision, any movement slower than ours, do you follow me?' 'Yes' - answered Eric - 'everything is clear so far'. 'Well, if this phenomenon has already happened and there is no actual reason to avoid it, there must be traces somewhere, or at least that was what I expected. So I remembered that Swan was working in Nigeria and asked her about the data and measurements she obtained from the underground. You know I am Brazilian, right? I am from the city of Fortaleza and I know all the area up to Natal very well. Well, I asked some colleagues at the University of Natal who work on Geology to give me data similar to Swan's. You know why?

Because the only way to find a rule within chaos is to find coincidences, which are never such'- said Oelio. 'It's like singing bossa nova. The instruments and voices are at certain points in the song, the rest is chaos!'

'We all studied this at school. Continents were formed by the subdivision of the Earth's crust, and it has been proven that Brazil's belly once coincided with Africa's groove. Especially the cities of Lagos in Nigeria and Fortaleza in Brazil used to be a few kilometers away. Not the cities, of course, but the places where they are now'. Eric looked at Oelio with a severe glance for the first time, like saying, do you think it is necessary to clarify that?

'Well' - said Oelio pretending he had not understood that - 'if it has already happened, it may have happened when Fortaleza and Lagos were close, and then, with the help of some physics at university, we compared results and found a rule within chaos. There are signs of all this, and the fact that the cities have, let's say, separated slowly, allowed us not only to date but also to find the frequency for this phenomenon: every 100 years.'

'Believe it or not, it is a question of alignment of the stars that surround us. The conditions for this to happen are given every hundred years, at least according to the astrologists and, well, also a couple of astronomers I know' - said Oelio smiling.

'The position of the moons affects the tides. The same happens with the shape of our ionosphere: the position of the stars changes its shape and allows it to tune and resonate at different frequencies'.

'Every 100 years' - said Eric - 'and when would be the last time it happened?' 'According to the calculations, in 1930' - answered Oelio.

Chapter 10 - The Prediction

Eric thought that, whatever Oelio understood by 'disastrous consequences' they would occur in a few months, there were less than two years left until 2030!

What had been a faint ray of light of dawn, was now vigorous broad daylight. It was 11 in the morning. Oelio was eager to explain the rest to Eric, so they took a pause to have some coffee. They had both understood that pauses are fundamental in the human learning process. Our brain is not a computer and it needs time to organize, clean, polish and put the new information in the right place. Rest enlightens the mind and also allows you to stand on a chair and look at things from a different perspective, then come down and continue.

This glissando, to resume the example of the finger on the piano, going from the lowest to the highest notes, did not produce any effects in the past. It has stopped in our atmosphere, crossing the planet with electromagnetic radiation of ever higher frequency until it disappeared. This phenomenon, according to our model, could last many days. A phenomenon that has been unknown until now, also because it is not visible to the human eye and the measuring instruments of old times, but nonetheless real. 'As I was saying' - continued Oelio - 'the conditions of the context have changed. It's like the forgotten cutting board for the lemon, like the glass cover of the panel in front of Saint John's church.'

'At the end of World War II, all the technology developed for military reasons was gradually released to society. Arpanet became the Internet, so now we have computers, semiconductors, software updates, and all that has entered our daily movements. If you think about it, almost everything we do and use has internal technology and silicon, right? The concept of computers has even changed. Information technology is inside blenders, washing machines, cars, factories, everywhere.

During the last century, silicon was unknown, or at least its characteristics; it was actually just 'sand'. Instead, this last century has been the century of silicon, much more than oil or other raw materials. Well, a shower of modulation frequency of 2Hz would have enough energy to go through everything and everyone bringing the components to high frequency in materials normally opaque to these frequencies. Like they say: don't put the phone near the induction hob or you'll roast it? This is what would happen: your mobile phone or at least all of its memory and information would be replaced by zeros (or maybe ones, I still don't know)'- concluded Oelio.

Now, Eric's glance had transformed with that fearful expression he had seen in Oelio's eyes until then. Eric was a technician, and it did not take him long to understand that the phone thing was just an example. Any technology saving rewritable information would be reset.

Every computer would be affected, like a planetary virus capable of reaching everything, even switched-off devices! There would be no escape, no alternative, no place on Earth to be safe from all this. And no place in space either, because waves, as Schumann resonance reminds us, are at the same time conductive and antennae. Have you ever wondered how FM radios work in tunnels? With a perforated cable, capable of transporting the signal in a continuous and coherent manner throughout the whole stretch. 'Yes' - said Eric - 'but because the same thing should happen outside the ionosphere: radiation should be contained within'.

'It will be our fault when it happens' - concluded Oelio - 'because we had an intact cable and we perforated it, it ended up like a tunnel. The ozone hole will allow part of the radiation to spread outside, rendering the space surrounding the Earth also unsafe from the 'virus'. Of course, thanks to the hole, the phenomenon will last less than the almost 30 days of the previous centuries, but enough to erase every kind of information saved in the planet. Every factory, every system will stop. There won't be any electricity, gas, Internet or water. Now everything is controlled by software. No device, not even the ones that work with solar panels, could work. They will be completely intact but will simply not work because the firmware (which would be factory software for the basic functions of a device) will no longer be there. Erased. It will be erased forever.'

Eric and Oelio remained in silence, lying on the sofas in the living room, each of them immersed in their thoughts. Eric thought about every time he had that feeling of emptiness when using technology and started to name things. He started to empty that overflowing bin, almost liberating, he thought.

All the knowledge of humanity erased in a few days, no programable equipment working and no way to save data, not even saving them in optic media because nobody would be able to read them afterwards. No place to hide things, not even the deepest cenotes would be able to serve this purpose.

His mind started to go back. History always calmed him. He thought how many times it had happened without anyone noticing. What if it had happened before, meaning, what if an advanced civilization before us had already lived this? Just like a computer, his brain remembered a meeting held when he often visited alternative scientific communities. He had heard quite a strange and fantastic interpretation of the Mayan prediction. He had laughed at them at that moment, but now the numbers bounced in his head.

About 2012, the guy said it was an incorrect reading because, according to Mayan mathematics, the number was not 2012 but 201+2, which is 203. However, according to these theories, Mayans divided their periods in decades, not years, so 203 would become 2030.

Chapter 11 - Mark

Mark had become the president almost as a joke. At the beginning, it seemed like a political marketing move, a sort of provocation. Then, with time, people started to understand it was not a game, or at least it may not be. It was a good idea for many, science fiction for others. The thing is, in the dawn of 2024, he had become the president of what once had been considered the most powerful nation in the world but, after the devastating Trump era, had inexorably descended many positions. A whole nation driven by a strong wish to be reborn. His first presidency was a good one, no false illusions, as some had predicted, but a man who had assumed command of a still very strong and big nation. He was running again for the 2028 elections, he was convinced that he could still give a lot to the country. His Facebook was now a distant memory. It was no longer mentioned, it was implicit. It had replaced the word 'Internet'. The web in itself no longer existed, it had also been replaced by Facebook. All the activity in the net had been incorporated into the 'network', as the detractors called it, making the classic inverted commas gesture with their fingers. However, when Mark Zuckerberg ran in the election again for a second term, they did not oppose either. Everyone thought he was the best candidate, even conservatives, and, even though he ran by himself, they supported his candidacy.

The 'Second Zuckerberg Era' had started the best way, with a plebiscite followed by quite a prosperous period. Industries in the United States were in full recovery, and the world had recognized his country's leadership again. Mark may have not accepted his candidacy if he had known what his administration and the whole planet would soon face.

The future does not lack a sense of irony and maybe now, more than on other occasions, this phrase would have that sinister meaning again.

Mark was sailing towards the second semester of his term. He was satisfied with himself; life had given him emotions few had felt before him. Everyone knows his story, his mistakes, his medals, the good and the bad Facebook has given to our society. However, it was an irreplaceable instrument now. It was impossible to think about the world of today, what it would have become, without the network, without social networks. Everything, from politics to economics to finance, from art to science and even religions revolved around the 'net'. Most traditional jobs had been replaced, sometimes by robots, sometimes by multinational companies that operated globally. Stores had disappeared. Those who did not work directly with information technology did it indirectly. Those who did not work either way were unemployed, period. Even politicians did all of their moves online. Elections were through the net, and all the administrative activities were somehow accessed exclusively through the network. The world as we knew it at the beginning of the millennium no longer existed. The last industrial revolution happened very fast. So fast, many realized when it had already ended and were confined to the periphery of society. The term 'periphery' itself had taken on a different meaning. It was no longer related to a place; periphery was social status. For some time, the concept of city center had vanished; and work, any work, was almost never linked to the place of production. Monday morning traffic had been replaced by an unreal desert, an artificial but evidently real Zabriskie Point. People left the house only after work to run, to relax. Expressions like 'I'm going to run some errands' had inevitably disappeared from the daily lexicon. Mailmen, street cleaners, gas stations, truck drivers, pharmacists, bank employees (bank owners had resisted), and so on had practically disappeared. Non voluminous goods were transported by drones. Everything was so automatic that people had forgotten their address; it was rarely necessary to give it. Ordering anything was enough to have it delivered to your address, wherever you were. The few remaining trades that were not related to technology could be counted on one hand: waiters, cooks, caregivers and nannies. Riding the wave of terrorism, the previous administrations had practically reduced the concept of privacy to mere jurisprudence, nothing more. The world had globalized, and almost every country in the world went by the same rules. Pockets of passive resistance to this model lurked in some areas of the emerging world: South America, Mediterranean countries, some areas of Asia and, of course, Africa, the eternally forgotten continent or maybe, as History would teach us, fortunately excluded of the modernization process. These areas of the world had turned into a dunghill of old technology, which was also feared but now necessary for their development.

This 'new world order' had turned 90% of the world population into income earning units. Teamwork was a distant memory; unions had almost ceased to exist. Everyone contributed for the system to work, just like a computer network, where every peer contributes to the final result. This similarity had been used by the rulers themselves to compare modern society with the human brain, formed by interconnected neurons. The way the brain worked had been known for a long time, as well has what intelligence, creativity and even disconformity were. What made the difference between people was not the amount of obtained information but the number of relations and connections in the brain. Like the game we all played as kids when our father left a magazine on the sofa: connect the dots. What mattered was not so much the final drawing but the dots and lines.

Leonardo Da Vinci, maybe before many others had been the representation of a man with a 'multiform intelligence' who taught humanity art, science and, above all, creativity.

Creativity might have been the only real treasure to discover in people, the only contribution that remained completely human and differentiated each dot of the drawing. Eric had read many books about Leonardo, the genius, and with the typical determination of engineers, had followed his teachings carefully. When he had to choose a faculty, he did not doubt it: Engineering. He would have even chosen it only for its root word, 'ingenium', which has so much to do with the Florentine genius.

Mark also admirer Leonardo, to the point that his enigmatic glance (which someone with a lack of tact would have defined as between 'I didn't understand a thing' and 'I am looking at the finger') was apparently a kind of "homage to Leonardo's mask". The thing is they had never met, although Mark, due to his origin, had direct contact with a great part of the world's scientific and academic world. Eric was too atypical to end up in a conference with the future president, too marginal to return to the big leagues.

Chapter 12 - Mother Nature

Perhaps not the world but Oelio, Eric and Swan were never the same from that night. When he came back home in the afternoon, Eric landed on the couch when he entered his house, even though his comfortable bedroom was a few meters away. He fell asleep and woke up at night, he left the house, took his bike, and started pedaling along the coastal road. The sea breeze in his lungs made him feel alive. He was not sure he was happy, but he was alive, a feeling he had not had in a long time. A kind of positive 'blank page panic'. A wish to do things, an energy that turned into kilometers and kilometers of coastline. His bicycle brushed against the road and he felt happy, he felt he was part of a great drawing, a huge fractal made of beaches, coasts, gulfs, inlets. Things he had seen thousands of times but now had a different meaning, a sign that Mother Nature never does anything by accident.

His mind started foreseeing apocalyptic results. The world could not stand such a blow. Everything humanity had built in the last 60 years would disappear or, in Oelio's words resonating in his mind: 'every technological device would lose its soul'. They would be intact but would not work. There was not an area of society that would not collapse with such an event. What would happen to people? The worst pestilences, wars, disasters would have been nothing in comparison. The number of deaths caused by the greatest conflicts in the world looked like a decimal number by comparison.

Humanity had replaced everything with technology, technology had saved data in a tiny electric charge in a piece of silicon. Attention centered on software viruses; the only ones capable of overcoming physical barriers for data protection. When, in previous centuries, information was transmitted in writing on paper, fire was the enemy. It was the virus that could have destroyed all human knowledge. We thought we had saved everything in a safe place and, on the contrary we had prepared the world for the worst disaster of all time.

'We should inform everyone' - he repeated to himself while he pedaled, 'the world should know'. One post would be enough to unleash an avalanche that would strike everything and everyone. How would the world react if everybody found out instantly? A tower of Babel 2.0! No, it was not feasible. It was necessary to act calmly, intelligently and cold-bloodedly. Looking for a connection in the Pentagon? Visit universities? He had no idea, he felt teeny tiny. This time, yes, just tiny. Overwhelmed by his thoughts, he stopped the bike, he lay on the sand and fell asleep in the fetal position. Contact with the ground made him feel safe, Mother Nature does not do anything by accident.

Oelio and Swan stayed connected through video for another hour. They had known each other for a long time and, when it was time for him to sleep, a new day started for Swan. The sun was already high in Lagos's sky; and when she was under the sunlight, she thought of Africa and saw it as she had never seen it before.

Everything that, at the beginning, seemed old, worn down, outdated, obsolete and rusty now seemed like precious relics of a past era. Precious because most of them would continue working even after the electromagnetic radiation shower.

Africa, the continent that had been forgotten, humiliated, subjugated and used as the planet's dumpster, had resisted the transition better than others, not only due to the different level of technology but mostly due to people's habit of not trusting it. This would not have saved everyone, but most. No doubt. And thus, 'the first world', as they referred ironically to North America and Europe in the countries of South America, would implode on itself. He felt he was living the script of one of those disaster films like Independence Day. Only there were no aliens or, rather, it was worse, much worse. The enemy is among us. Our enemy is ourselves? Were the Wachowsky sisters right when they said humanity was a virus that had to be destroyed? Was it the way Mother Nature had guaranteed life on the planet, wiping out most of the human race with the end of technology? Destroying factories, nuclear plants, oil wells but also pollution, overexploitation, desertification...

As Oelio said, the true legacy we are leaving for posterity is plastic, the whole planet is full of plastic and soon microplastics will enter our diet, and Mother Nature will not have a remedy for this, or will she?

Swan started his day at work with these thoughts. For her, this was year zero, day one.

Chapter 13 - The First Step

National Security Advisor Alfred Hostings had been requested by the President in person. Former security specialist of the larger information technology multinational companies, he had a brief past in the Mossad, and, after that, he decided to resume teaching in Munich, the city of his parents. Mark called him on the phone one day after his reelection. They had met and work and respected each other very much, even though they were from completely opposite social backgrounds and often disagreed. Mark told him he was his 'mirror of truth', one of the few who, in an unsuspected time, had called him a jerk without hesitating and, for him, that deserved a medal. Mark was like that. Even though he was evidently no longer a teenager who was full of pimples and crazy about Asian women, he had managed to preserve some aspects of his character. What some considered 'instability' or 'volatility' was 'intuition' or 'genius' for others. Whatever his talent, it had helped him in the past, and would also do it in the immediate future.

Oelio had not met Hostings in person, but as any good Brazilian knows, there is always a distant relative somewhere in Brazil. It is an unwritten rule that describes well the country and its colorful ethnic mix, which has given place to great cultural and artistic moments, as well as possibilities.

If you could read the names on the doors of the houses in a Brazilian city, you would find a kind of condominium Esperanto, a multiplicity of cultures in close contact.

His neighbors always told him about this cousin, a grandchild of emigrants who left everything to go to the Americas after the Second World War. His last name was Haustin, but alterations were common among immigrants' last names.

Oelio called his mother and asked her to put him on the phone with her neighbor Genny Haustin (actually Gertrud).

At that moment, it seemed to him like the best way to reach out, nothing official, no devices, just two neighbors from far away (very far away).

He had no idea how to speak to Hostings, but he obtained the Advisor's mother's email, and immediately started working.

He asked for a picture of his mother and Genny, attached it to the e-mail and wrote that he had talked to the neighbor, who remembered she had met them, and he played with Alfred when he was a kid. He wrote to her saying he had old pictures and wanted to send them to him, but he did not know how to contact such an important man.

The woman answered after a few hours, happy with the picture and anxious to also see the pictures of her little boy, who was now a man. 'Here it is. This is the e-mail we use to contact him. He begged me not to give it to anyone but the family, but I think I can make an exception for you. Don't forget to send them to me, too',

she answered with the linguistic precision only a German can render. That mixture of icy coldness and kindness of heart, typical of the southern Germans .

Oelio finally had a contact, finally, he thought. That wasn't very hard. I thought it would be more difficult. Now what? Tackle the issue immediately or work around the topic? Should he introduce himself as a slightly crazy scholar or an old neighbor? He decided not to do any of those things. He wrote to him forwarding his mother's picture and explained him that the pictures he was talking about were in a box labeled by his grandmother 'for Alfred', he did not think it was appropriate to just scan them and he preferred giving them to him in person.

Alfred, slightly surprised by so much attention, accepted and invited him to his office in Washington. Eric and Swan were immediately informed about Oelio's progress. Eric insisted on going too, but in the end they decided it would be more prudent and less suspicious letting Oelio give the first step with Hostings. His funny crazy mathematician appearance would help him get out of trouble, as it had happened in the past.

The plane landed at 1 pm, Washington time. In less than half an hour Oelio had arrived at the city center, at National Security Advisor Alfred Hostings's office. Oelio had an accurate plan: he untidied his shirt, sprinkled himself with water as if he was sweating and entered Hostings's office.

Alfred Hostings stood up and received him with a welcoming gesture, a typically Brazilian greeting gesture, offered him coffee and they sat in the peach-colored office. Oelio started talking to him about Fortaleza, about them as kids, his grandmother, his mother, all of his loved ones, trying to avoid talking about the photographs. Alfred got carried away but Oelio's velvet voice and almost forgets about the photographs. At that moment, Oelio said: 'By the way, I left the pictures at the hotel. If you want, I can go and get them'. Alfred answered they could be sent the next day. He was not surprised by his interlocutor's forgetfulness. He was not surprised that he had been chatting with a perfect stranger who was actually giving him pictures for a very specific purpose. Oelio was a mathematician but, above all, he was a teacher. He knew when to lull his students to sleep and when to wake them up, it just depended on the voice tone. He asked Hostings a random question, waited for him to finish his answer politely but without listening to absolutely anything he was saying, and then he added: 'my dear, I could have also scanned them, but unfortunately I stopped doing that'. Then, a silence.

Alfred almost woke up with this unusual silence of such a loquacious interlocutor. He looked at him and couldn't but ask '¿Why?'.

Finally, the question Oelio was waiting for. Oelio poured himself another cup of coffee and told him everything, starting with 'Because I prefer paper, it lasts more'.

Chapter 14 - The Team

When Oelio finished talking, Alfred lit a cigarette nervously, looked at Oelio in the eyes and said, 'We had never seen each other and the pictures don't exist, do they?'. Oelio nodded and shrugged. 'What else could I do?' - he answered - 'What were the odds that one of you listened to me?'. Hostings had enough experience and knowledge to understand that the person in front of him was not crazy. He remembered the Mossad and the subterranean experiments with very long waves, they might have achieved that as well?

I can organize a meeting next week with my team of experts. As a Security Advisor it is my duty and, in my heart, I hope that what you are telling me is a load of crap. That is how he said goodbye. Oelio closed the door and left.

The following week, Oelio, Eric and Swan were received by a team of experts designated by Hostings. He wasn't there. Scott Forstall, former Apple employee specialized in operative systems was presiding over the meeting. The reunion lasted 3 hours; Swan, Oelio and Eric talked in turns following a script with accuracy. At the end of the meeting, they were asked to leave a copy of the documentation. They would be contacted as soon as possible. Forstall's dark gaze when they said goodbye froze Eric.

Ten days went by with no answer. It was certainly not appropriate to contact Hostings, and Oelio had the idea to send a flower bouquet to Munich, to Hosting's mother, with a car that said 'I apologize but the pictures were lost with the baggage. I send you a hug and thank you for helping me'. She replied via email, she said she liked the gesture very much and that her little boy was with her and that he was sending Oelio a warm greeting.

The following week, at 7 am, a van stopped in front of Oelio's house. Agents with civilian clothes got out of it and called at the door. Good morning, we apologize for the time, but they have asked us to pick you up and take you to a safe place. You have 10 minutes to pack and come with us. If you have any complaint, you can make it later. It is 7:04 am, at 7:14 am, whether you like it or not, you'll be on the back seat.

Oelio was petrified. Never had such hard words been pronounced so politely. He would have never thought

English was capable of such dichotomies. Eric and Swan had the same luck. At 9:15, the three of them met in a waiting room in an unknown location. The windows of the vans were completely dark. Mobile phones didn't even work in the passenger compartment, which indicated that the vehicle was more than armored. So much diligence after more than two weeks and then, why that 'safe place'? Safe from whom, others or themselves? They waited nervously for 10 minutes. After that, they took them to another room. It was obvious that the facility was subterraneous, or at least that is what it seemed, without telephones, windows or external noise.

The great table in the middle of the room was a disappointment for Eric. He was expected a great oval table made of finely carved wood, microphones, ergonomic chairs, fresh water in steel glasses. However, he found an aseptic room with a big table made of 10 desks placed one next to the other. There was no sign of technology, nothing suggesting that the President himself would enter afterwards.

They were invited to seat down. There were about 10 people, among which they recognized Hostings and Forstall, the others were perfect strangers to them. After two interminable minutes of silence, a door opened and Mark entered alone, without an escort or anyone to open the door for him. He entered, sat down and started to talk.

'Ladies and gentlemen, we are gathered here to evaluate the impact of XELF on the planet. In particular, its effect on our technology. With us here, professor Oelio, engineer Eric and researcher Swan. Preliminary conversations have already taken place with Alfred and Scott'.

The three of them though: 'What an unusual thing, at least he should say our last names, it's not like we are at school'. It was Mark's modus operandi. He had learned that last names are annoying, bulky and anti-communicative. It is better to use first names and simplify the language. It is easier even for those who don't speak English well and, above all, it preserves the anonymity of the attendees.

'XELF?' - asked Eric - 'what does it mean?'. 'It is an acronym that was coined to describe that hypothetical phenomenon you have revealed', answered a voice from behind. 'Who am I talking to?' - asked Eric. 'I am Larry' - he answered politely. At that moment, Eric recognized Larry Page, who had invented Google and his heart skipped a beat. He looked at

Oelio and Swan to make sure they had already recognized Page, and then he remained silent.

Mark invited the participants to start the discussion and stayed in a corner to observe them. The most brilliant minds of the last 50 years were in that table, some of them known, others not so much, but each of them with a narrow specialization in their sector. There were geologists, physicists, biologists, mathematicians, computer technicians, engineers, philosophers, historians... A table where anyone would have been able to make any question and would have always received the right answer.

Eric and the others had an initial moment of due shyness and then, helped by the absence of surnames, they started behaving normally. The meeting lasted approximately three hours, after which Mark, with great skill, managed to silence everyone and remind them that they would meet again the following morning. Eric, Swan and Oelio looked at each other, and they turned to Alfred at the same time. He told them: 'We have prepared your accommodation. Your clothes and belongings have already been brought here. We told your neighbors that you had travelled for work. We also told your employers, relatives, etc.

Outside this door, our staff will guide you to your accommodation. See you tomorrow'. The three of them tried to rebut this when Mark stood up, looked at them and said: 'As President, I must protect everyone's safety and avoid information leaks that, at this stage, may be extremely dangerous. I hope that, as citizens, you realize that this is a matter of national or actually planetary security'. He nodded in greeting and left. The three of them had time to digest these words and realized they were alone. They went towards the door Hostings had mentioned, a bit like sheep that return quietly to the pen in the evening.

Chapter 15 - Isolation

Eric, Swan and Oelio had been accommodated in a great apartment next to the meeting room. It was a very comfortable 300-square-meter apartment. They had a radio and a television, no Internet but a little spa, three comfortable bedrooms, each one with its own bathroom, a small meeting room, a perfectly equipped kitchen and a living room. The apartment had windows, but they were sealed. Outside, there was a perfectly maintained and guarded garden. Everything was made so that the light entered without being able to look at the sky and, consequently, the stars. There was no way to get oriented, except deducing the cardinal points from the movement of the sun. The staff was extremely polite and, in general, they couldn't complain. They had free access to the pantry, which made Oelio extremely happy, and a full menu available 24 hours a day. A place where you could stay for a long time without too much trouble. This last aspect froze Swan, would they be isolated for a long or short time and in relation to what?

Tired by the hustle and bustle, the news and the thousand questions in their brain, they all sought refuge in something. Eric, as usual, landed on the bed and stayed there for a couple of hours. Oelio inspected the kitchen and, above all, the pantry, while he eased his mood with a snack, while Swan immersed herself in the bath and, afterwards, in a restorative nap.

It was a little after 8 pm. They met in the living room, which was already covered in Oelio's notes, with crumbs scattered everywhere. 'Well, what do you think?', - exclaimed Oelio, looking up and discovering a ketchup stain on the collar of his shirt. Swan, compressed in her leggings and a black bra that made her look like a sadomasochist, a kind of female version of 50 shades of Gray, got up, looked at the colorful garden and took a deep breath.

'I am not sure I understood the situation, but I would say we have attended a secret presidential meeting in a bunker that has nothing military or governmental about it. They said they would take us to a safe place but, as there is no imminent danger, I would say the place is not safe for us but safe from us, don't you think?' 'Exactly' - answered Eric - 'I agree. They are scared we might disclose the news. They isolated us because we were the only ones in the room without a government office, the only ones who hadn't sworn, the only ones to control. This apartment must be full of microphones and cameras, someone must be looking at us right now and transcribing everything we say. Did you write the word 'transcribing' well? Careful, it's difficult!' - exclaimed Eric, looking defiantly at the ceiling. 'I recognized about 50% of the attendees' - added Oelio - 'and, as far as I know, it was the elite of science and technology. Such a high-level team and a place like this mean that the danger is even more real than we think'. 'Actually - he added - the mathematical model is quite optimistic.' 'Optimistic?' - repeated Swan - 'define optimistic, please'. 'Let's say the duration of the resonance could also have an effect for a month. In that case, any hope of saving information is null. A month of these things and the world will become a great blank page'.

The light suddenly went down and the three of them clearly heard a dark sound, almost a vibration, more than a sound. One single hit, firm and isolated. 'A storm is coming' -

exclaimed Eric. Swan was petrified, her blood had frozen, a feeling of anxiety went through her heart like a sword. He recognized that noise, it was very similar to the sound of a shale oil extraction plant. He said nothing to his companions. Dinner was delivered at 10 pm sharp and, for a moment, radiations, disasters and questions led to a cheerful silence interrupted by the sound of Oelio's chewing, and Eric and Swan laughing with their mouths closed.

After all, they were three human beings who loved each other and had learned to appreciate each other. Even Eric and Oelio, despite the fact that they had met recently, were very close. Theirs wasn't a game of three, they were three different relationships, each one with its own rules, actors and script, practically a family.

A few hundred meters away, in one of the bunker's laboratories, it was broad daylight. Twenty-three technicians, experts and military personnel worked intensely on a new type of simulator. A great sphere of about 6 meters in diameter seemed to be floating in the middle of the room, and people in the lab moved like ants in the anthill. Every one of them had something to do, few words, lots of gestures and many concentrated but worried looks, this could be seen at a glance.

At 11 pm, Swan, Eric and Oelio said goodbye and each went to their room. Eric left the door open, as usual. Instead, Oelio and Swan locked themselves in their respective bedrooms. The following day, they were sure, would be long and exhausting, so it was better to sleep.

Before falling asleep , Oelio noticed that, among the things the bunker personnel had brought from his house was a small wooden box he loved. 'Tomorrow I must ask how I could smoke!' he thought.

He fell asleep smiling, who would have said his good mood was so important for the government!

Chapter 16 - The Calculations

Oelio's astrologist had calculated well the influence of the stars to determine the phenomenon's start date: April 2030. The same result had been obtained by the President's personnel. Exactly a year before the world goes back a century. The stock market crash, the panic, the lack of the most essential services, the pestilence, millions of deaths. It was necessary to act with precaution and learn from history. History, which Eric loved so much, now seemed to be the only Rosetta stone capable of showing us a way to survive. Mark was frowning. All the good technology had done, all the wellbeing it had brought would be swept away in an instant and turn into the worst epidemics that human history would be able to remember or mention. It was him and nobody else who wanted to include Oelio, Swan and Eric in the team of experts in charge of finding an antidote or, at least some kind of solution. He trusted his instinct, he had done it before, and it was certainly not the time to make distinctions. Anyone who would have brought a ray of hope was welcome. The team's task was to verify the veracity of the prediction with experimental evidence. The following morning, everyone met at nine o'clock in the meeting room. Except Mark, everyone was present.

'Good morning. As you can imagine, Mark won't be present in every meeting, but the President is constantly informed about our work' - said Alfred. 'What would be our job, if I may ask?' - continued Eric. 'Eric - answered Alfred - this team's task is to verify with factual evidence that Schumann resonance at a very low frequency and its effects really pose a threat to the planet, and to find a solution.' 'Factual evidence?' - said Swan taking the floor and capturing everyone's attention as usual.- 'How do you plan to find facts that confirm this? If only it was theoretically feasible to build a tool to prove this, we probably wouldn't be here' 'And where?' - answered Alfred — 'someplace in the world selling the solution to the highest bidder?' Swan was speechless, she wasn't sure she had understood correctly, and, in any case, she was not expecting such bluntness from a government official. Oelio took the floor: 'I think that not only this affirmation is unfair, but it is also out of place. Swan, Eric and I never hesitated to contact you. I speak for myself, but also for them. We are here due to our civic sense'. 'Let's keep calm' - said a woman named Marissa - 'Alfred, I don't think that is the right tone for this situation'. Alfred took the floor. Oelio, Eric and Swan were surprised by Marissa's imperative tone. He said: 'I apologize, these are very difficult days and I have personal problems, let's try to start again'.

Alfred looked at Oelio directly in the eyes and whispered, 'my mother is not well'. Oelio realized those were the eyes of 'her boy' for that woman he might have met as a boy on the stairs of his house. He was profoundly moved and answered Alfred with a smile full of "saudade", like only a Brazilian can do. 'I'd like to inform our guests' - he said in an official tone, making his scared puppy look disappear, and then he turned to Eric and Swan - 'that last night, it was not a meteorological phenomenon that caused electricity to drop, and what you heard was not thunder'. 'I bet there were spyholes and cameras' thought Eric. 'You have been accommodated in a government facility. Your existence is known by the participants in this meeting, yourselves, the staff and, of course, the President. 'Yes'- he added, smiling -'you are here so that the outside world is a safe place. I apologize for this and I understand it may be unpleasant to hear it, we owe you an apology but, as the President says, it is a matter of planetary security. We cannot risk releasing this news uncontrollably because it would only make the situation worse'.

All signs of distrust had disappeared from Eric, Swan and Oelio's eyes, they were thankful to Alfred for speaking so frankly again.

'In this facility's laboratories, we study new technologies. This is not, I repeat, a military facility. These are civilian facilities requested by the President himself to study new technologies.

A reverse satellite transmission system has been studied for a long time. Before talking, let me finish'.- he said with a hand gesture, interrupting Eric, who looked like an Olympic athlete who was ready to run the final 100-meter dash -'We were studying a system to transmit signals at planetary scale without using satellites which, as you know, have already invaded all the space available around the Earth and turned it into a dumpster hanging over our heads. An inverse satellite system is a system that uses the Earth itself as an antenna to transmit the signal toward the Earth's surface. Imagine the uses of this technology; antennae wouldn't be necessary; for the first time, the signal would "come out" directly from the ground. By telling you it is a transmission system based on the use of very low frequency waves, not only have I broken a dozen federal acts of law, but I may have also answered most of your questions'.

Alfred made a pause to make sure there were no more questions, and he continued: 'In the laboratory, we are trying to recreate at a very small scale, the conditions to activate Schumann radiation at 1.5 Hz. We have been working on that since the day after our meeting' - he said looking at Oelio.

'The first meeting, the one about the photographs?' 'Yes' - answered Alfred letting out a smile. 'So not everything was bullshit' - said Oelio smiling - 'Not really' – interrupted Marissa. 'Your predictions, your model, are consistent with ours. You are missing a few data to produce an accurate forecast but, all things considered, it is a great job'. 'Unfortunately, the new model is much less nice than yours' - said Larry, while he stood up. 'According to the new forecasts, the resonance will take place on April 28next year, and it will last 23 days, more or less. All known storage devices, RAM, EPROM, EEPROM, ROM, etc. will be reset to zero. Yes, even ROMs will be erased, it seems like science fiction! We are looking for alternatives but, at the moment, we have nothing. All modern technology is based on digital electronics. No device will be safe. All analogic circuits will survive, your grandfather's stereo will keep on working, as long as it has a record player and records to listen to. Old cathode ray tube televisions, old radio receivers, some washing machines, but everything else will be erased, eliminated during the first 24 hours. We are saving data in optical media, but we won't be able to read them for a long time'.

'At the moment - interrupted the oldest attendee, named Bill - all the governments in the world are preparing similar rescue plans. From the next 24 hours, the price of paper will double every day, according to our analysts' predictions. A UN meeting has been scheduled for next week, including all the non-member countries who wish to participate. A closed-door meeting to discuss the next steps'- concluded Bill with a final sigh.

Chapter 17 - United Nations

The United Nations building may have not ever seen such full participation. Practically all nations in the world were present, including North Korea, Cuba and the Islamic State (which, as well as the Spanish ETA, had disarmed in 2020; some said due to Trump's merit; others due to history's inevitability). You could cut the air with a knife. Many had big bags full of paper, photographs, diagrams. At the center, there were stenographers, all of them older than seventy, pressing the keys with skill and also a little arthrosis. Many of the participants had never seen a stenographer in their lives and asked themselves why they had those strange supermarket cash registers in the middle of the room.

The closed-door meeting lasted four hours. Almost half of the participants knew nothing about the imminent danger. It wasn't easy for Mark to inform about the certainty of disaster and the impossibility to somehow stop it. Every effort was centered on saving as much as possible. Except using the Earth's soil as insulator, the only possibility was to send a very fast ship to space in the hopes of insulating it enough to preserve its precious content. Technically, a "Faraday cage". Practically, an ark. World heritage in a small expedition to space, which would land on Earth ten months later as long as navigation systems, which are necessarily digital, had resisted. Mark, the new Noah, explained that a great flood of electromagnetic radiation was about to hit the planet and, just like the meteorite to the dinosaurs but in a much less spectacular manner, it could have wiped out the human race or, at least, society as we know at least for the last hundred years. Every Chief of State would have to warn about this the following week. In seven days, at 12 Eastern Time, the whole world would know this bitter truth. They would know that meteorites, pestilence or much less wars had not endangered humanity. There hadn't been a third world war but, as Einstein had predicted, we would return to the cave. This time, the enemy would remain quiet, invisible, but no less cruel. In a few months, the world would have to prepare to a disaster never seen before. People would know one of the darkest ages of humanity. A Middle Age where mafias would take control, as they are used to controlling people's territory and needs. Big cities would implode under the weight of people's needs. Small towns, and especially the fields and isolated areas would suffer less. The more one depended on services, the more one implicitly depended on technology. Water, electricity and gas supplies would be inevitably interrupted. Of course, there was time to reestablish the production cycle, to 'migrate' electricity lines, gas ducts and water pipes to last-century technologies, but time would certainly not be enough for everyone and, above all, user demand in 2030 could not be satisfied with technologies that were more than 100 years old.

One June 28, 2029, the world would find out about the terrible truth. Obviously, an information strategy was developed, and words were chosen, as well as the tone and what information and advice to give humanity. At the end of the meeting, each participant received a package labeled "Top Secret" and a small cardboard box. Each package had been translated to each first language, and it contained data, statistics and all kinds of information about the event.

At the end of the meeting, the delegations of the different countries did not issue any declaration. There were attempts to leak information, but they failed miserably thanks to the generalized control the governments already had over the net. They all returned home with the difficult task of organizing the information campaign, the measures to avoid riots and panic and, above all, filling the insignificant 1000 Terabytes offered by the UN for all that was intended to be saved in the ship. Almost every nation in the world reorganized the army which, with the years, had often been replaced by drones and robots, they stocked printer paper and looked for as much technological material (televisions, radios, high-frequency devices, etc.) of the analogical type as possible. It was decades old, but it could work after the imminent planetary disaster.

Seven days, less than 170 hours. Many asked themselves why the rush, others tried to extend the deadline. All the member countries with veto power, maybe for the first time in the history of humanity, agreed to alert the population as soon as possible according to an agreed schedule.

Eric, Oelio and Swan agreed with this decision, also because it would bring them closer to the day they could come out of the bunker and resume their normal lives, whatever that phrase would mean later.

A week of meetings and studies determined again the inevitability of the event, but, to Oelio's surprise, the harmful effect of the ozone hole had been reduced by transporting the electromagnetic storm to the space surrounding the Earth. The idea of the ship had been developed by the team, included its maximum load and, consequently, given the technology of the time, the number of terabytes reserved for each country. The ship would have to reach enough distance to protect its load for six months and, after a break of many weeks, return to Earth while everyone waited for it like a stork before a delivery. Carrying out that mission in such short time, two months, required an unprecedented economic effort and an exchange of military technology that would have never been revealed without this urgency. So in a short time, members of the whole humanity, industries included, were working on this new Ark. This time, there were no animals to save or rain to avoid. The ship would not have a crew and would not be able to report its status to the Earth once the storm had passed. The world would remain upside down for about ten months, waiting for a sign from the sky, waiting for a messenger capable of giving back what the storm had snatched from it.

Humanity had evolved a lot since its beginnings, it had learned about democracy, religion, war and technology. Once again, the future did not lack a sense of irony.

Millenia had passed for us to find ourselves, like ancient humans, with the eyes in the sky waiting for a sign.

Chapter 18 - G4

On June 28, 2029, exactly ten months before the disaster, from the Oval Room of the White House, President Zuckerberg and, at the same time, 48 Chiefs of State read United Nation's official declaration. A very strong CME (coronal mass ejection, a solar geomagnetic storm) would affect the planet. As opposed to the previous ones, the level of this one would be G4 (in a scale of G1 to G5) and the most common electronic devices would stop working. The great industrial structures, data centers and all the archive of the 'net' would be protected by the existing safety infrastructure. Do-it-yourself attempts to protect equipment were not recommended. The recommended method was to upload one's own data to the net (documents, photos, videos) to preserve them. As an alternative, common paper printings or other physical supports could be another guarantee to save them.

They did everything possible to calm down the population, 'governments have been acting for some time, and utilities, stock exchange, banks, nuclear plants, gas, etc. were not at risk'; it was all false!

All the Chiefs of State committed perjury. Nobody would ever know, no one would ever be believed. That was the strategy, period. The world could not afford a hecatomb before the disaster.

Phrases like 'the most common electronic devices' did not alert multitudes, but many started imagining apocalyptic scenarios. As in the best traditions, the population was divided between catastrophic and not, between those who understood, intuited, that behind those words, hid a very different truth. They all thought about electronic organizers, telephones, they all thought about 'obviously' digital devices. Few suspected that, even limiting the disaster to home appliances, it would render 'soulless' almost all surrounding technology. That it was not about not being able to access the net or call someone. It would be a suicide of all that technology they called normality. Those who had pacemakers underwent surgery, when possible, to replace them with devices that were not digital.

Soulless, that expression Oelio used was probably the one that best described the aftermath. Common use objects, sometimes necessary, would be deprived of life.

Oelio himself did not manage to go beyond those words. For many members of the team, despite their level of education, it was hard to imagine an aftermath. Somehow, many of the attendees waited for a desperate gesture of humanity instead of a last-minute technical solution.

The team was dissolved. Eric, Oelio and Swan returned (under oath) to their lives. They would keep in touch with the government during the next few months, but it was time to face the outside world. It was time to get off the roller coaster avoiding violent reactions and dizziness. The three of them were being watched by the secret service, as in the best tradition. Keeping watch over people had become extremely simple: a bit, a 1 instead of a 0 and your life became a reality show for the employees. They could find out anything: how many times a day you used the washing machine, which programs you watched on television and, of course, communications. Nothing you did during the day lacked a trace. The relevant authorities knew everything. Eric and the others knew it well. They were taken with the same technique that had been used at the beginning: dark windows and no chance to locate the bunker. They had managed to guess that it was within a radius of about 200 km of Eric's house.

Going back to normal couldn't be harder. Perhaps it was simply impossible.

The following morning, they got up early (they had all stayed at Eric's house), they preferred to stay together. Despite the forced stay of 4 weeks in the bunker, they still needed to be together. They still needed each other, maybe because among millions of people, they were the only ones they could talk to.

April 28, 2030. They all knew that day; they all knew an electromagnetic storm would transform the ionosphere that day, as well as most of the Earth's crust in a wave guide. An electromagnetic wind would sweep it all off. A gigantic eraser would eliminate all traces of our lives, our past and, who knows, maybe our future.

The requests to discourage home-made solutions were useless. Not so much because they were dangerous but because they were useless. That absolute certainty of the governments that individuals were unable to provide a solution gave many people the idea that things may not be exactly like that something may have been silenced. It did not discourage those who started digging very deep holes in the ground in the hopes of preserving their things. Some buried telephones, computers, memory banks. There were also televisions, refrigerators and any digital home appliance. During the months that followed the announcement, the world seemed to have rediscovered agriculture. Everyone was obsessed with the impending disaster and many, as children do with their freshly fallen milk teeth, buried their belongings as if they could sprout the next year. 'In millennia, they will find strange plastic urns with sand inside, silicon will slowly turn into what it once was. In contrast, and sadly, plastic will last much longer' - commented Eric when he read this news. All the work was useless. There was much preoccupation with preserving things and not much willingness to create a lifestyle in accordance with the imminent new era. Had humanity lost that survival spirit that had helped it in the past? Actually, technology had not only paved for the greatest disaster history remembers, but it had also transformed the majority of people into perfect victims of this new plague, unable to come up with a practical solution, only capable of analyzing it. Synthesis, not analysis, repeated Oelio. We have taken care of teaching people how to analyze things but not synthetize them. We have voluntarily forgotten about the creative process underlying our evolution.

Perhaps it was not like that, or perhaps it was. One thing was certain: humanity would not be ready, and the number after the letter 'G' would 'calm them' much more than the astonishing truth. The electromagnetic storms would be at a G6 level (meaning 'beyond our measuring capabilities').

Chapter 19 - The New Cuba

The arrival of 2030 had been different from other years, with few celebrations. The world knew that number '2030' would be remembered by future generations. Other historic dates, such as the discovery of America, the Second World War, the attack on the Twin Towers, would be secondary event compared to the year 2030. The signs were already evident. Old cars, bikes and scooters, gas heaters and gas tanks had already reappeared. In January 2030, gas tanks price was already ten times higher and, even though they were mass-produced, the demand was enormous. Paper had also started to grow scarce, while, for water, there were cases of entirely walled rooms transformed into water tanks.

Humanity was giving an answer, and it wasn't precisely the smartest one. Nobody knew which one was the right one, but one thing was certain, this one wasn't it.

Near to the fatidic date, Oelio, Eric and Swan met. They all decided to meet at Swan's house in Lagos. Africa, with its backwardness and instability was, paradoxically, the safest place.

The city of Lagos had turned into the capital of the 'third world'. In the 20's, large emigrations from the North of Africa had two very specific destinations. Those who had attended school and had a specialization tried to go to Europe. In contrast, those who belonged to the less privileged class, and consequently were less educated, without a specialization or knowledge, went to Lagos. Its population had quadrupled in one decade, and it may have turned into the area of the world with more workforce and poverty. In a globalized world, Lagos was unique. Maybe this singularity would make the difference. That's what Oelio thought. He knew about monetary disasters in South American countries: Brazil, Argentina, Venezuela. They had already been through at least one default and they had survived thanks to the arms and brains of the people, certainly not thanks to diplomas.

'In moments like this, the best thing is to be with those who have always lived in balance', said Swan.

Was Lagos the new Cuba? After the Castro era, Cuba had turned into conquered ground again, and all the human material narrated during the end of last century had disappeared. Old American cars, symbols of the islands, maintained by those who had been recognized as the best mechanics in the world for decades, Cubans, had disappeared. Now, also in Cuba, anything could be bought online, including cars.

Yes, Lagos was the new Havana and Equatorial Africa was the new Cuba. All obsolete technology was there, the best mechanics were now there, even globalization counterculture was there, a small focus of nonconformism in an ever connected and plain world. While the world went crazy waiting for the fatidic date, people there moved on. Perhaps they were not trying to answer the question. Maybe they already had the answer, they just had to wait a few weeks to also have the question.

Naturally, Nigeria also participated uploading data to the ship, known as 'The big zip', which had left a long time ago without a fuss. They tried their best not to cause more concern, and the sheep was launched from Cape Canaveral a cold November morning. A brief press communication informed the world population that The Big Zip had left for his U-shaped course. This technological boomerang would come back to Earth on December 23, 2030, an artificial comet would bring the soul of technology back to Earth as it had been 'frozen' a few months before.

The last days were the worst, or at least that was the common belief. People had compulsive deliria; many went to the bank every day to verify everything was working. Printer noises were heard almost everywhere. Walking at night down the empty streets of Paris, Rome, Buenos Aires, Tokyo or Washington, you could almost only hear the sound of printers. The lights always on, even at night, almost to exorcize the darkness, the technological Middle Ages that would soon come. Everyone tried to save something, everyone tried to print memories, experiences, moments, events and anything that had been photographed or filmed. It was a bit like filling an imaginary backpack with things. The movies that marked their lives, the books that accompanied them, the meetings that changed them, everything in a digital support should be printed or transformed into an analogical format. The record and movie market were reborn. Record players, especially those with reloading cranks, were sold like hot cakes, just like slide and film projectors. They all had a very good reason to convert life events to an analogical format. They all felt the need to break that imaginary line that was being traced. Oelio, Swan and Eric made their own list of 'things to save'. It was like a game, they sat around a table listing everything that had moved them in their lives or even just brushed them.

On April 27, 2030 everyone was silent, ready for the event. Those who had to do something had already done it. Those who had to move had already moved. Whatever you had to do; you would have probably done it already. It was a not-day, a kind of last day of the year without fireworks (at list visible to the naked eye). Everyone asked themselves how long it would last and how long it would take to erase everything.

The following morning, the measurement instruments developed by the governments of the whole world started to identify an atypical electromagnetic radiation of growing energy everywhere in the world.

Chapter 20 - Spring Cleaning

The world is divided between those who love and those who hate midseason. It is almost like a political stance, at a certain moment in life we decide if we are on the side of immobility or on the side of transition. Whether we like it or not, whether it is Spring or Fall, transition seasons are a sign of novelty, and we often expect something without realizing. It was Spring in the Northern hemisphere and Fall in the Southern hemisphere. In any case, a transition season, it couldn't be any other way.

The color of the sky, at least in Lagos, was intensely blue. Nothing made you think of the disaster that was occurring in the whole world. The effects started a few hours later. The resonance reached full 'maturity' after a few hours. For many, the experience was similar to the death of a loved one. They held their computers or phones in their hands as if they were newborns they had to protect. Screens froze and, after a while, as exhaling their last breath, they remained on. Then, the darkness.

Everything that was powered by electricity, solar power or batteries went out. All uninterrupted power sources also went out. The digital electronics that made them work also went out. Loading fuel would have been useless, not even generators worked, only the oldest ones.

Every factory, car, train, subway, all means of transportation went out. Planes, the few of them who dared fly that same day, went into free fall, like birds shot by an invisible hunter. The eleventh plague of Egypt, for some. Mother Nature's answer, which had been expected for a century, to the systematic destruction of the planet perpetrated by humanity, for others. All forms of communication that were not direct stopped. Nobody knew anything anymore and many left the house as if it was the first time and started to walk.

The big cities were no longer deserted. People left their houses massively, ignoring the martial law, and without the distractions of the modern age. After a few hours of desperation, they did the right thing: they opened the windows, which were permanently closed due to the air conditioning, and looked at the world with new eyes. They went out, talked to random people, any stranger seemed like a brother, like someone you knew you loved but hadn't seen in a long, long time. There were tears, but also laughter, hugs; the other had been discovered. The neighbor was no longer an apartment number, he was one of us, made of flesh and blood, with his defects and virtues. It was that strange guy living upstairs we didn't know what he was doing at home all day. It was the neighbor in front of us; we considered her strange only because he lived more during the night than during the day. We only knew those useless things about the 'others'. We didn't know anything true, personal certain. Everything was 'I heard..', 'I think...', 'I'm sure...'. Now it was not the time for selfies, likes or messenger groups. Now it was the time to get our hands dirty and talk to the other one. There was no 'Enter' key and, above all, there was no 'Undo' key. Words had weight, just like looking at each other in the eye when speaking, just like shaking somebody's hand to congratulate him or hugging him to cheer him up. There were no emojis to describe our own emotions with a catalogue. There was reality, like thee had been for centuries before us and maybe again after the storm.

Life changed for everyone from one day to the other. Cash lost value in small towns and in the countryside. In the big cities, it remained for some time, but was quickly replaced by bartering.

Everything had vanished. The most fortunate people had been left with an apple or a droid on the screen. Many only received a cold and very enigmatic message on the screen: 'press any key to continue'.

Today, fifty years after the disaster, we don't know if the ship has returned or whether its valuable contents were saved or not. Communications are nonexistent, as well as long trips. News from the rest of the world arrive in fragments. Life continues here in Lagos, but it seems like the population could not stand the strike in other parts of the world. Plagues, wars and desperation rule most of the world. We receive new from other still viable places like ours in South America and remote regions of Asia.

Eric, Oelio and Swan told me this story. I was just a teenager, but they made sure they told me the story, each of them from their point of view. They instructed me about what to do and how to transmit it to posterity. A thousand copies of this story in two thousand PET bottles, the small ones within the big ones. The last one, full of water.

Today, the world is not what it used to be. Society as I knew it when I was a child no longer exists. It imploded on itself.

I don't know if humanity will resist. I don't know if we will also go extinct like the dinosaurs and maybe other civilizations before ours.

I live in Lagos and I feel lucky to be here, who would know.

Chapter 21 - At School

Thanks Zucky for reading us The Text, great job. Certainly, some words cannot be completely replaced with our vocabulary, but, as it was written almost 2000 years ago, your job has been really remarkable.

Now, be quiet and go back to your seats. Snorting is useless, you'll just blow your candles and, as you know, wasting fire is against the law.

So, a quick summary of the text we just studied. It has been found in several samples within the sacred containers of the PET dynasty. Several concepts have not been clarified by the experts yet, but this story is thought to have been written anonymously on a support that is similar to paper, with a strange acronym: 'Homero'. We don't know the meaning of this acronym.

In the text, the word 'technology' appears many times. It is the name of a benevolent goddess who brought abundance to all the prehistoric population.

The fierce Schumann tried to overthrow the power and take over the world but the son of God, Zuckerberg, took care of all humanity by trying to protect it. This is, in a few words, the official interpretation given by the priests.

We are here because there is a second, unofficial and forbidden interpretation that states that the PET were aliens who came to our planet millennia ago in flying vessels. The sacred containers had been built by the Evian, who had settled in our lands like the Coke, the Pepsi and the Perrier, prehistoric tribes in charge of releasing this text in sacred containers. These containers are made of an indestructible material, available in many parts of the planet in the most varied shapes and colors. They called it plastic. It is transparent, odorless, and apparently, all the PET world was based on this substance, in addition to, of course, sand. The latter was found in many urns (of various shapes, mostly rectangular, and numerous sizes) containing a very thin, dark sand, also available in many parts of the planet. We don't know the use or the reason for this sand contained in the urns. USB was engraved directly on the urn, on the outside.

There are many rhetorical figures in the text. Do not take it literally, try to understand its meaning, the story and what the author really meant to say.

You have four hours to complete the task, the time you have until the candles go out.

And now, keep silence.

There are two 'apparent' errors, rhetorical figures or inconsistencies that anticipate part of the finale, if you have not discovered them go to: pressanykey.it

Index:

Everyone wonders, but who is the author?

Uncle Enzo, a communist and godfather at my christening, demanded a Russian name. My mother, a primary school teacher, wanted a short name. Thus, Sandro was chosen instead of Alessandro.

Since April 11, 1964, I have been learning the concept of .zip.

An electronic engineer seduced by Informatics; I started my career as a radio host long before I started losing my hair.
I like South America, Roger Rabbit, awkwardness among strangers in an elevator, limescale remover, looking more like my father as the years go by, Music (only good), staying in silence with a friend, ownerless fields in the middle of roundabouts and everything superfluous.

I am one of those who would be called a nerd in slang, who just opens a computer and does everything. One of those you meet for dinner at a friend's house and vomit all your problems on, with the cell phone, computer, alarm system, washing machine with Wi-Fi, air conditioner with Wi-Fi, thermostat with Wi-Fi, lock with Wi-Fi. And, of course, the one you ask how to hide your messages and pictures, because 'you like dealing with these things, it's your passion, isn't it?

I invent show titles (hits), even if I still haven't received a dime. I am witty, nice and... oh, yes, solar!

I wrote a book, I mean, a novel. A short novel, actually. In fact, I did not write it but printed it on paper directly from my little head (size 62).

www.ingramcontent.com/pod-product-compliance
Lightning Source LLC
Chambersburg PA
CBHW030702220526
45463CB00005B/1867